Rogério Tomanini

Genetic factors that influence psychopathy and sociopathy

Rogério Tomanini

Genetic factors that influence psychopathy and sociopathy

Genetic Sociopathy

ScienciaScripts

Imprint

Any brand names and product names mentioned in this book are subject to trademark, brand or patent protection and are trademarks or registered trademarks of their respective holders. The use of brand names, product names, common names, trade names, product descriptions etc. even without a particular marking in this work is in no way to be construed to mean that such names may be regarded as unrestricted in respect of trademark and brand protection legislation and could thus be used by anyone.

Cover image: www.ingimage.com

This book is a translation from the original published under ISBN 978-613-9-66781-9.

Publisher:
Sciencia Scripts
is a trademark of
Dodo Books Indian Ocean Ltd. and OmniScriptum S.R.L publishing group

120 High Road, East Finchley, London, N2 9ED, United Kingdom
Str. Armeneasca 28/1, office 1, Chisinau MD-2012, Republic of Moldova, Europe
Printed at: see last page
ISBN: 978-620-8-05804-3

I dedicate this work to my mother, who has always been by my side at all times, advising me and helping me on my journey, and to my wife Gabriela Arbex Buono Tomanini, who with great wisdom and patience has accompanied me on this new journey.

The scorpion approached the frog on the riverbank. As he couldn't swim, he asked for a lift to get to the other bank. Suspicious, the frog replied: "Come on, scorpion, only if I was too foolish! You're treacherous, you'll sting me, release your poison and I'll die. Even so, the scorpion insisted, with the logical argument that if he stung the frog they would both die. With promises that he could rest easy, the frog gave in, put the scorpion on his back and started swimming. At the end of the journey, the scorpion stung the frog with its deadly stinger and jumped unharmed onto dry land. Hit by the poison and already beginning to sink, the desperate frog wanted to know why such cruelty. And the scorpion replied coldly:

"Because that's my nature"

SUMMARY

Summary

Since the formation of the concept of society, man has studied and dedicated himself to understanding human behaviour, even before the conventional social model found today existed. In ancient Greece, one of the most worrying human behaviours - violence - was already being studied. Alcmeon of Cretona (6th century BC) was the first to dissect animals and dedicate himself to studying the biopsychic qualities of delinquents, researching the human brain and looking for a correlation with their behaviour. It was said that in man there is a bit of animal and a bit of God, that life is the balance between the opposing forces that make up the human being and that illness corresponds to the disruption of this balance. One of the most notable men in history, Hippocrates, the "father of medicine", believed that all crime, as well as vice, was the result of madness, thus laying the foundations for imputability or the principle of criminal irresponsibility of the insane man. With the onset of the Middle Ages and throughout this period, crimes were linked directly to a great *peccatum* (sin), and were punished by severe measures of imprisonment, torture and the death penalty. It can be seen that the study of human criminal behaviour has been perpetuated over the centuries in a solid way, in an attempt to explain the reason or reasons for criminal acts and what would be the effective measures to stop such acts, applying punishments or treatments. In the study of criminals, one of the most notable researchers of the contemporary era who linked criminals to their physical characteristics was Cesare Lombroso, an Italian university professor and criminologist, born on 6 November 1835 in Verona. He became world famous for his studies and theories in the field of characterology, or the relationship between physical and mental characteristics, relating physical structures such as jaw size to criminal psychopathology, or the innate tendency of sociopathic individuals to criminal behaviour. Advances in technology and research have shown that human behaviour is directly related to physiological and, nowadays, genetic characteristics. Based on this precept, this study proposes an analysis of the endogenous factors that lead human beings to criminal behaviour, tracing an organic pathway for certain reactions, analysing the origin of certain cellular and molecular dysfunctions. It proposes a broad discussion of human behaviour from a physiological and genetic point of view, looking at two interconnected aspects: the organism and the environment.

Keywords: Criminology, Genetics, Human Behaviour.

CHAPTER 1

Introduction

The approach of Césare Lombroso (a scholar of human behaviour) is a direct descendant of phrenology, created by the German physicist Franz Joseph Gall at the beginning of the 19th century and closely related to other fields of characterology and physiognomy (the study of mental properties based on an individual's physiognomy). Lombroso's main idea was partly inspired by genetic and evolutionary studies at the end of the 19th century, proposing that certain criminals had physical evidence of a hereditary "atavism" (reappearance of characteristics that were only presented in distant ancestors), reminiscent of more primitive stages of human evolution. These anomalies, called "Stigmata" by Lombroso, could be expressed in terms of abnormal shapes or dimensions of the skull and jaw, asymmetries of the face, etc., but also of other parts of the body. Subsequently, these associations were found to be highly inconsistent or completely non-existent and theories based on the environmental cause of criminality became dominant.

Parallel to Lombroso, other researchers were endeavouring to develop theories to explain the criminal act of human beings, including Enrico Ferri (who classified criminals into five types, namely: born, insane, occasional, habitual and passionate), Raphael Garófalo (creator of the term Criminology, elaborating his conception of natural offence starting from the Lombrosian idea of the born criminal) and Adolphe Quetelet (creator of Scientific Statistics, Based on three principles, he established the so-called Quetelet Thermal Laws), he tried to demonstrate that in winter there are more crimes against property, in summer there are more crimes against the person and in spring there are more crimes against customs (due to the exacerbation of sexual activity that occurs at the beginning of this season).

These scientists, among others of great importance, although with divergent theories, dedicated themselves to understanding the criminal behaviour of human beings, looking to the environment and the individual for reasons or motivations for antisocial and anti-legal practices.

With the advance of technology in all fields of science, it has been possible to develop more in-depth study techniques to understand the functioning and individual characteristics of each human being. More specifically in the field of medicine, technological developments have been of great value in diagnosing illnesses and seeking new forms of treatment.

The study of molecular biology, through new technologies, made a substantial leap forward in the 20th and 21st centuries, as never before since its creation. With the discovery of D.N.A. (deoxyribonucleic acid) by the German biochemist Johann Friedrich Miescher (1844 - 1895) and the subsequent explanation of its functioning and components by Watson and Crick in 1953, who won the Nobel Prize for the discovery, studies into human anatomy evolved to considerable levels. However, the great leap forward in the evolution of molecular biology at experimental levels began with the Genome Project (1990), which aimed to map the human genome, with the expectation of identifying approximately 100,000 genes. After this mapping was completed, it was concluded that the human being has around 20,000 to 25,000 genes, generating approximately 400,000 proteins.

Following these discoveries, a race began in all directions of the health sciences to identify and treat all types of diseases that affect human beings, with the basis of the studies not only being the effects of the diseases, but also the molecular causes of these ailments.

[a]In the field of forensic psychiatry, two diagnostic manuals are used, the *Diagnostic and Statistical Manual,* currently in its 4th edition (DSM-IV) published by the American Psychiatric Association (APA) and the *International Classification of Diseases*, published by the World Health Organisation.

(WHO), identify a personality disorder sharing similar characteristics (with certain divergences). The DSM-IV classifies it as antisocial personality disorder (Axis II, Cluster B) while the corresponding diagnosis in the ICD-10 is dissocial personality disorder. However, some authors argue that these criteria do not go far enough to define a third entity called **"psychopathy".** These blurred lines of classification, disagreement among mental health professionals, poor understanding of the biological and non-biological (environmental) factors precipitating and maintaining this behaviour, add to the confusion.

CHAPTER 2

BACKGROUND

The study of human behaviour, due to genetic characteristics, is a relatively new science, which takes us back to the study begun by Cesare Lombroso. However, molecular biology, together with other fields of science such as sociology, psychiatry and anthropology, are coming together to understand and resolve human behavioural issues through each specialisation. Given the complexity of the subject, it is necessary to delve deeper into the relationships between human behaviour and the biological effects on that behaviour.

CHAPTER 3

OBJECTIVE

3.1 General Objective

To evaluate the relationship between criminal and anti-social behaviour and hormonal, molecular and genetic factors in humans.

3.2 Specific Objective

Analysis of endogenous factors related to hormones and genes that interfere directly and indirectly in the behaviour of individuals who display antisocial and aggressive behaviour.

To see what the legal and punitive consequences are for individuals characterised by criminal pathologies that are directly related to genetic dysfunctions.

CHAPTER 4

Genetics

The role of genetics in determining violence and aggressive behaviour has recently been examined. In addition to the possible interaction with hormones (testosterone, serotonin and corticoids), it has also been surmised that the *MAOA* gene polymorphism has an interactive association with childhood adversity in predicting aggression in men. This observation has been repeated in several studies and offers an interesting example of a possible interaction between genetics and environmental factors. Analyses of single nucleotide polymorphisms (SNPs) in a sample of adolescents with antisocial behaviour and drug addiction reported significant genetic associations for two genes, *CHRNA2* and *0PRM1*, compared to controls. The former gene codes for the neuronal nicotinic receptor a-2 (associated with nicotinic dependence in schizophrenic families) and the latter for the opioid receptor p (implicated in many drug abuse behaviours). Similar findings for a genetic connection in a dual diagnosis of substance abuse and conduct disorder symptoms have been reported. They showed evidence of linkage to the 9q34 chromosomal region when both vulnerability to drug addiction and conduct disorder symptoms were considered. There was also evidence of linkage to the 17ql2 region for conduct disorder symptoms alone. The evidence from the twin and adoption studies shows that both heredity and environment have the same influence on antisocial behaviour. However, a later analysis showed that the influence of heredity is greater in children with antisocial behaviour, who are more insensitive and emotionless, compared to the control group (CHATURAKA et al., 2010).

Aggressiveness, lack of emotion and insensitivity are not merely the result of environmental factors. Biology has an equal part to play. Evidence about the neurocircuitry of empathy and callousness has emerged in recent years. This system has a complex relationship with the neuroendocrine system through control and feedback mechanisms. A state of neuroendocrine imbalance (lower activity in paralytic structures and hypoactivity of the hypothalamic-pituitary-adrenal axis to stressful situations) contributes to insensitivity and lack of emotion, which can self-perpetuate over time (CHATURAKA et al., 2010).

CHAPTER 5

Hormones

Hormones are chemical messengers, facilitators of various functions necessary for human survival, participating in body mobilisation in the face of danger, the search for rewards such as food or sexual partners, and social skills. They also affect the ability to learn after punishment or reward and the tendency to take risks. As phenotypes, i.e. intermediate biological mechanisms at the molecular level that link genes to the manifestations of mental disorders, hormones, when deregulated, can contribute to the symptomatology seen in psychopathy. They represent viable biological markers for a variety of scenarios and investigations, and may find agreement with the versatility of the clinical presentation of psychopaths (BARROS et al., 2015).

Some studies indicate that psychopathy can result from an imbalance in the levels of both cortisol and testosterone, particularly through the relationship of increased testosterone and reduced cortisol. This relationship results from mutual inhibition between the HPA (hypothalamic-pituitary-adrenal) and HPG (hypothalamic-pituitary-gonadal) axes. The amygdala is a brain region affected by this asymmetry, as it is an important binding site for both hormones. Low cortisol levels (reflected in reduced fear) and high testosterone levels (uninhibited behaviour and reward-seeking) can modify the responsiveness of the amygdala, reducing sensitivity to punishment or fearful stimuli. Hormonal imbalance involving a decrease in cortisol and an increase in testosterone can jeopardise connectivity between subcortical regions (limbic system) and cortical structures. Neuroimaging scans of adult and juvenile psychopaths suggest that connectivity between the amygdala and prefrontal regions is compromised, affecting the decision-making process, as emotion-related information from the amygdala that signals danger does not reach cortical areas to inform decisions. Uncoupling can also decrease the ability of cortical regions to send inhibitory signals to subcortical regions, resulting in deficits in emotion regulation and behavioural inhibition and contributing to emotional instability and the reactive forms of aggression found in psychopaths (BARROS et al, 2015).

5.1 Serotonin

Serotonin or 5-hydroxytryptamine is a naturally occurring neurotransmitter, especially in the human brain, responsible for conducting nerve impulses. 5-HT is an indolamine resulting from the hydroxylation and carboxylation of the amino acid L-tryptophan. The first step in its synthesis in the central nervous system (CNS) and in other areas of the body, such as enterochromaffin cells found in the intestinal mucosa, platelets and mast cells, is the uptake of tryptophan. Tryptophan, which in turn comes especially from the protein diet, is actively transported by carriers common to other amino acid chains. Thus, the level of tryptophan, especially in the brain, is influenced not only by its concentration in the blood plasma, but also by the concentration in the blood plasma of other amino acids that compete for these same protein carriers. Tryptophan undergoes the action of the enzyme tryptophan hydroxylase, passing into the form of L-5 hydroxytryptophan. This latter form is then transformed into serotonin by the action of L-amine acid decarboxylase. This enzyme is highly distributed and has a wide range of specificity to different substrates, which makes it practically impossible to control serotonin levels in the brain via this enzymatic pathway. This neurotransmitter fulfils its role through interaction with various receptors. Based on their structural and operational characteristics, 5-HT receptors are subdivided into seven different classes (5-HT$_i$, to 5-HT$_7$), with 14 subtypes identified. The 5-HT$_{1A}$ receptor acts as a somatodendritic auto-receptor responsible for modulating the activity of serotoninergic neurons. Apparently, activation of this receptor modulates emotional and eating behaviour, cognitive functions, maturation and cell differentiation. The 5-HT$_{1B}$ and 5-HT$_{1D}$ receptors modulate the release of serotonin and other neurotransmitters, such as acetylcholine. 5-HT receptors$_2$ are linked to the visual cortex, modulation of eating behaviour and mediation of vasoconstriction, while the 5-HT receptor$_3$ is responsible for modulating the release of 5-HT and is apparently related to mechanisms of pain perception, release of acetylcholine and dopamine, as well as gastric motility and secretion of enteric fluids. Serotonin is a vasoactive amine that acts on the cardiovascular system, smooth muscle and promotes platelet aggregation, not to mention that it acts as a neurotransmitter in the CNS, being especially related to the limbic system, controlling the reactions of: anxiety, fear, depression, sleep and pain perception (http://www.infoescola.com/neurologia/serotonina/ 2017).

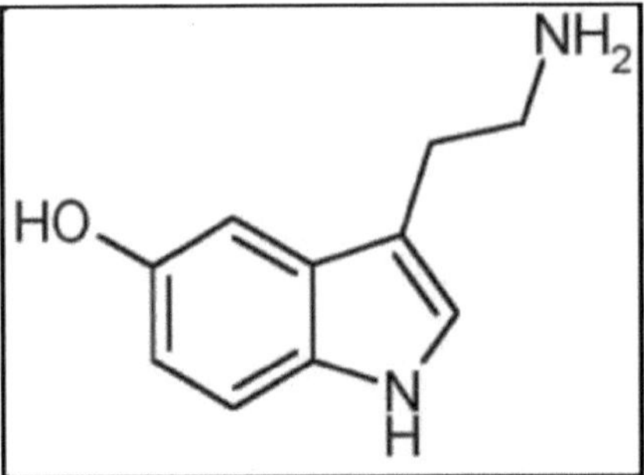

Figure 1 - Serotonin molecule. Serotonin is a neurotransmitter, i.e. a molecule involved in communication between brain cells (neurons). It is chemically represented by 5-hydroxytryptamine (5-HT) and is also often referred to by this name.

Source: (Department of Education - Government of Paraná, 2017).

Extracellular changes in the levels of this neurotransmitter occur in different behavioural states. A decrease in serotonin levels is known to increase sensitivity to pain, exploratory behaviour, locomotor activity and aggressive and sexual behaviour. In both humans and animals, mental disorders have been correlated with alterations in serotonin function, such as aggressive and obsessive behaviour, as well as attention deficit. In relation to sexual behaviour, this neurotransmitter plays an inhibitory role on the hypothalamic release of gonadotropins, with a consequent drop in sexual response. However, the decrease in serotonergic activity makes sexual behaviour easier (http://www.infoescola.com/neurologia/serotonina/ 2017).

Two of the most studied molecules involved in regulating serotonin levels in the brain are the serotonin transporter (5HTT, also known as SLC6A4), which transports serotonin from the extracellular space, and monoamine oxidase A (MAOA), a key enzyme responsible for serotonin degradation. Both genes encoding these proteins harbour genetic polymorphisms in their promoter regions that have been shown, *in vitro,* to affect their transcriptional activity. These genetic variants have been studied for their possible involvement in a range of psychiatric conditions and behavioural characteristics. The serotonin transporter polymorphism (5HTTLPR) and the monoamine oxidase A polymorphism (MAOA-LPR) have been implicated in conditions and behaviours such as depression, anxiety, aggression, alcoholism, autism, suicidal tendencies and impulsivity (NORDQUIST; ORELAND. 2010).

The brain structures involved in emotional regulation form the neuronal basis for how we respond to external stimuli. Research has shown that the phenotypic outcome is modulated through an interaction between environmental insult and genetic factors. In a sample of male adolescents, antisocial behaviour was shown to be more common among carriers of the low-functioning MAOA-LPR variant, since they had also suffered childhood abuse. Similar results were observed in depression, where carriers of the low-functioning 5HTTLPR variant were particularly vulnerable to stressful life events, demonstrating that the penetration of a genetic factor in behaviour seems to be dependent on the environmental context (NORDQUIST; ORELAND. 2010).

5.2 Testosterone

Testosterone is a sex hormone secreted by the hypothalamic-pituitary-gonadal (HPG) axis. It is linked to psychopathy because its levels are much higher in men than in women, and may account for the higher prevalence of persistent antisocial personality disorder in males (10 to 14 times more prevalent in men than in women). Certain psychopathic traits, such as reward-seeking, dominance and aggression, are associated with testosterone. High levels of this hormone have been observed in girls and boys with conduct disorders, juvenile delinquents and female criminals. In addition, testosterone has been associated with difficulties at work, non-compliance with the law, drug use and alcohol abuse. A direct link between testosterone and psychopathic traits has not yet been established, but evidence suggests that this hormone interacts with others, predisposing to psychopathy (BARROS et ah, 2015).

53 Corticoids

Cortisol is the hormone released by the hypothalamic-pituitary-adrenal (HPA) axis, the most studied of all the endocrine axes. It is released in response to a stressor and potentiates the state of fear, generating sensitivity to punishment and promoting withdrawal behaviour - areas in which psychopaths show deficiencies. When a stressful event occurs, signals from the limbic system (amygdala) and regions of the cerebral cortex trigger the pituitary secretion of corticotrophin-releasing factor (CRF) into the bloodstream. CRF stimulates the release of adrenocorticotrophic hormone (ACTH) by the adenohypophysis, whose function is to mobilise body resources and provide energy in times of stress (BARROS et al., 2015).

Psychopaths show reduced stress responsiveness, fearlessness and low amygdala functioning, leading to the hypothesis of low cortisol levels in these individuals. Low resting cortisol levels have been associated with impaired fear reactivity in young children, increased sensation-seeking in men and greater financial risk-taking. Psychopathic offenders also exhibit lower cortisol levels than non-psychopathic criminals (BARROS et al., 2015).

5.4 Dopamine

3-hydroxytryptamine, or dopamine, is an important neurotransmitter involved in motor control, endocrine functions, cognition, compensation and emotionality. This biogenic amine plays an important role in memory and cognition. In addition to its role as a neurotransmitter in the CNS (Central Nervous System), dopamine acts as an inhibitory transmitter in the carotid body and sympathetic ganglia. It also appears to act on the distinct peripheral dopaminergic system. It induces various responses not attributable to stimulation of classical adrenergic receptors: suppresses aldosterone release; directly stimulates renal sodium excretion; suppresses noradrenaline release in sympathetic endings by an inhibitory presynaptic mechanism; relaxes the lower oesophageal sphincter, delays gastric emptying; causes dilation of the renal and mesenteric arterial circulation; regulates the activity of cholinergic intemeurons in the striatum and the release of acetylcholine; participates in the regulation of haemodynamics and electrolyte transport, as well as in the secretion of renin (ESTEVINHO; FORTUNATO, 2003).

As far as the CNS is concerned, dopamine is present in 4 main pathways:

1) Mesencephalon (more specifically the substantia nigra) to the involuntary motor areas of the basal nuclei (striatum); deterioration of the cells in this area leads to Parkinson's disease; 2) Mesencephalon to the frontal lobes; these pathways appear to be related to attention and orientation and may be involved in drug addiction and hyperactivity leading to attention deficit; 3) Mesencephalon to the limbic system (control of emotional responses); these areas appear to be related to the reinforcement and stimulation centres and may explain the dependence on drugs that increase dopaminergic function; they also include areas that appear to be hyperactive in schizophrenia, which explains why drugs that block the effect of dopamine are used in the treatment of this pathology; 4) short pathway related to the release of hormones from the pituitary gland (ESTEVINHO; FORTUNATO, 2003).

Many aspects of how the human brain performs the calculations essential for healthy social interactions remain unclear. The superposition of neural representations of the value of one's own and others' outcomes is a central component of empathy and pro-social behaviour, suggesting that attention should be paid to the neural systems involved in evaluation. These include the neuromodulators serotonin and dopamine. In fact, many psychiatric disorders associated with monoaminergic abnormalities show social dysfunction and interactions between serotonin and dopamine have been implicated in impulsive aggression (CROCKETT et al., 2015).

A role for dopamine in harm aversion is less clear. Although biomarkers of hyperactive mesolimbic dopamine function in humans is correlated with aggression traits and impulsive antisocial psychopathy traits, direct evidence supporting a causal influence of dopamine on human antisocial behaviour is scarce. Previous studies have shown dopaminergic effects on economic decisions, but existing economic models are poor of moral decisions regarding harm to others (CROCKETT et al., 2015).

The dopaminergic system is involved in behavioural activation, motivated behaviour and reward processing. It also plays an active role in modulating aggressive behaviour. In animal studies, hyperactivity in the dopamine system is associated with increases in impulsive aggression. Studies on aggressive behaviour in rodents have shown that high levels of dopamine have been continuously observed before, during, and after aggressive fights. In humans, it has been linked to the recognition and experience of aggression. After administration of the Dopamine D2 receptor antagonist sulpiride, subjects showed a decreased ability to recognise facial expressions of anger. There is also evidence that impulsive behaviour can be increased by high dopaminergic function. Stimulants increase impulsivity in humans without the presence of a disorder, and pharmacologically manipulated dopamine levels have been shown to increase or decrease aggressive behaviour (SEO et al., 2008).

D2-targeting antipsychotic agents, dopamine receptors, reduce anger levels in aggressive patients. Agents such as risperidone, clozapine and olanzapine have also been effective in treating impulsive aggression. Overall, the evidence from these studies suggests that there is a substantial involvement of the dopamine system in aggressive behaviour (SEOetal.,2008).

5.5 Interaction between Serotonin and Dopamine

The serotonergic system has strong anatomical and functional interactions with the dopaminergic system, more specifically, a reciprocal interaction between these two systems. Behaviours related to approach and withdrawal are thought to be determined by the balance between dopamine and serotonin activity, with dopamine thought to stimulate appetite behaviours and serotonin thought to discourage appetitive behaviours and provoke their withdrawal. Interactions of this kind between the serotonin and dopamine systems provide a framework for understanding the mechanisms underlying impulsive aggression. Considering the functional regulation of serotonin over the dopamine system, impaired serotonergic function can result in hyperactivity of the dopamine system, promoting impulsive behaviour. This relationship may explain the co-occurrence of serotonin and dopamine dysfunction in individuals with impulsive aggression. In support of this, prefrontal serotonin levels in rats decreased to 80% of baseline during and after fights, while prefrontal dopamine levels increased to 120% after fights. These results suggest that the decrease in serotonergic activity in the context of aggressive behaviour is closely associated with the increase in dopamine activity. In line with the research, dopamine-triggered impulsivity in rats was increased by serotonin depression or removal of the serotonin-1B receptor gene. In humans, low levels of

5- HIAA and high levels of the dopamine metabolite homovenilic acid (HVA) were associated with high scores on the interpersonal and behavioural items of the Psychopathy Checklist (SEO et al., 2008).

CHAPTER 6

Genetic psychopathy

Between 1501 and 1596, there was one of the first medical descriptions of psychopathic personality, described by Girolano Cardamo, a professor of medicine at the University of Paiva. This doctor's son was beheaded for poisoning his wife. In this description, the doctor speaks of "improbity", a condition that does not reach complete insanity, because people still had the ability to direct their wills. After years of study, Canadian researcher and psychiatrist Robert D. Hare, developed the Psychopathy Checklist-Revised method in 1991, in which psychiatrists assign a scale of 0 to 2, based on a clinical assessment and the patient's personal history, to the following topics: good lip; inflated ego; unbridled lying; thirst for adrenaline; overreaction; impulsiveness; antisocial behaviour; lack of guilt; superficial feelings; lack of empathy; irresponsibility and childhood misconduct (BERTOLDI, 2013).

According to Eduardo Teixeira (Forensic Psychiatrist), research shows that criminal behaviour is related to the **HTR2B** gene (responsible for the production of Serotonin), which can predispose its carriers to impulsive attitudes. For the most part, this genetic inheritance is present in criminals, but it is important to note that the existence of this gene does not presage the individual's impulsive behaviour (BERTOLDI, 2013).

Studies show that recent stressful life events and childhood maltreatment predict depression in young adults in proportion to the number of "short" alleles (deletion) carried from a 44 base pair (bp) insertion/deletion (long/short) polymorphism in the regulatory region of the serotonin transporter gene (5-HTTLPR) (BYRD, MANUCK. 2014).

Men with higher levels of serotonin have greater impulse control in general, and particularly sexual impulse control, as well as lower levels of aggression and anxiety. If this hypothesis is correct, an important prediction would be to see low levels of this neurotransmitter in psychopaths. Mealey (1995, p. 531), in an exhaustive literature review, cites a series of studies in which precisely this relationship has been found; psychopaths, criminals and other individuals with high *scores* on measures of aggression and impulsivity have significantly lower levels of the serotonin metabolite 5-HIAA. These effects are not small, reaching an average sampling effect (difference between the high and low score groups divided by the standard deviation) of 0.75 (CALLEGARO. 2010).

In women, high levels of serotonin can initially increase the likelihood of accepting sex, as anxiety and resistance are reduced. However, with chronically high levels, such as under the effect of antidepressants, there is a reduction in sexual desire and inhibition of orgasm, sometimes even preventing the orgasmic response. Women with low serotonin levels, on the other hand, are more excitable and reach orgasm easily, as well as displaying more initiative and aggression, often taking the reins of the sexual relationship (CALLEGARO. 2010).

But how does this neural control of impulses work depending on the social context? The cerebral cortex is responsible for modulating impulses. The frontal lobes, and especially the prefrontal cortex, exert a decisive influence on the control of sexual or aggressive impulses, although the pathways involved are still poorly understood. The prefrontal cortex evaluates situations and makes decisions based on the context, and has been identified as responsible for the ethical management of our behaviour, partly due to its inhibitory capacity, delaying the gratification of impulses. It is precisely the prefrontal cortex that is implicated in psychopathic behaviour; patients with frontal lesions start to act impulsively, no longer controlling their sexual or aggressive impulses, as in the classic case of "Phinéas Gage", described by the neurologist Antonio Damásio (CALLEGARO. 2010).

CHAPTER 7

Syndromes

Syndromes are inappropriately considered by society to be diseases. In reality, they are genetic conditions characterised by different groups of pathologies to which the individual is exposed as a result of the same illness. Genetically caused syndromes are generated, as you might expect, by genetic factors, with physical and psychological manifestations due to various malformations, which vary according to the nature of the syndrome. There is no cure for this type of anomaly, since the syndromes develop with the carrier and can have causes related to chromosomal alterations or genetic mutations. Many genetic syndromes are detected at birth or in the first few months of life, while others begin to show their most characteristic manifestations in pre-adolescence or even adolescence, delaying diagnosis. Early detection of the problem is fundamental so that appropriate medication can be used, minimising physical, inconvenient and social damage, so that the individual can receive the necessary follow-up so that they can develop more closely to social standards, improving their quality of life (SYMPTOMS AND TIPS. 2017).

7.1 Klinefelter Syndrome (47,XXY)

Bobby Joe Long, violently raped and murdered at least nine women from May to November 1984 in the city of Tampo - Florida (USA). He was sentenced to death for raping fifty women and killing nine. He was born with a truly unusual condition known as **Klinefelter's Syndrome,** which means that he had an extra "X" chromosome (female) causing greater amounts of female hormones (oestrogen) in his system, with the physical consequence of growing breasts at puberty, which caused him great embarrassment (http ://truecrimecases.blogspot.com.br/2012/08/bobby-joe-long.html. 2017).

7.2 Supermasculinity Syndrome (47,XYY)

Sex chromosome trisomy 47, XYY is the most common male chromosomal aneuploidy compatible with live birth. It is estimated to have a high variability in living men, ranging from 26 to 375 per 100,000, although many are undiagnosed or diagnosed late. Trisomy 47, XYY is much more frequent when studied in a tall population, which is explained by the presence of additional copies of the SHOX gene (and possibly also other genes related to height) in men. The XYY sex chromosome abnormality has been described in various configurations since the first descriptions of a group of men with 47, XYY in 1965 by Jacobs *et al* who carried out a chromosome survey of male patients at the State Hospital in Carstairs, Scotland, where it was discovered that men with 47, XYY karyotype were particularly frequent among prisoners in penal institutions. During the 1960s and 1970s, studies of people with 47, XYY identified an increased frequency in hospitals for the mentally handicapped, and in prisons. Several of these studies reported a general increase in the rate of criminal behaviour and an increase in crime, especially due to sexual crimes. These studies were associated with selection problems as they investigated institutionalised individuals. Two relatively new studies of criminal behaviour among sex chromosome trisomies have been published. Gõtz *et al* found an increased rate of criminal behaviour among people with 47, XYY. One study indicated that people with 47, XYY had a four-fold increase in convictions, mainly due to offences. All the research carried out so far on this issue is limited by the study of selected groups, institutionalised patients or clinicians, as well as methodological shortcomings such as self-reporting of crimes, ill-defined definition of crime type and ill-defined control groups. All studies were also conducted on very small groups, including <20 people with a chromosomal abnormality. The full spectrum of all crime types has never been reported (http://astrissomias.blogspot.com.br/2011/05/trissomia- xyy.html. 2017).

CHAPTER 8

Genetic Sociopathy

The concept of a personality disorder with insensitivity (Sociopathy) as well as disrespect for social norms is clearly established in psychiatry. These people share a combination of traits that can include violence, aggression, insensitivity, lack of empathy and repeated acts of criminality against social norms. However, the classifications and definitions from this point on are unclear. Although these traits have existed in human societies since time immemorial, identifying and classifying such behaviour has changed over time and continues to do so. In fact, the understanding of personality and its disorders was quite different at the beginning of the 19th century from its current context (which refers to a collection of traits that are expected to have a biological basis). So the term "personality" was thought to be more of a metaphysical question. However, as the century progressed, the measurement of personality in more objective terms, and therefore the objective description of its disorders, gained popularity (CHATURAKA et al., 2010).

Poor regulation of the serotonergic neurotransmitter system (Serotonin) is directly linked to this behaviour. Serotonin is thought to help control aggression, impulsivity and disruption of this system, resulting in fewer restrictions. Indirect evidence for this hypothesis comes from the reduction of aggressive and impulsive behaviour with serotonin reuptake inhibitors (SSRIs) in normal people. However, attention should also be paid to recent criticisms of the presumed efficacy of SSRIs based on the theory of neurotransmitter imbalance. In animal models, reduced activity in the serotonergic system is associated with increased attacks against non-vulnerable targets (offensive aggression). Predatory aggression towards vulnerable targets was not affected. There is also evidence that the serotonergic system closely interacts with cortisol control and testosterone secretion. Disruption of the serotonin system is assumed to be partly responsible for the diminished cortisol response to stressors (CHATURAKA et al., 2010). An interesting association between testosterone and a functional polymorphism of monoamine oxidase A (*MAOA* **gene)** has been demonstrated. The underlying hypothesis is that testosterone has a direct effect on the transcription of the *MAOA* gene by acting on one of the promoters. However, the stimulation of transcription is not as strong as that of glucocorticoids, which also bind to the promoter. When testosterone levels are high, they can competitively inhibit glucocorticoid binding and result in less transcription of the gene. The gene's product, monoamine oxidase A, breaks down a multitude of amines, including serotonin. Using 95 male criminal alcoholics and 45 controls, it was shown that a combination of high cerebrospinal fluid testosterone level and a low *MAOA* genotype activity were significantly predictive of antisocial behaviour and aggression in men (CHATURAKA et al., 2010).

CHAPTER 9

Limbic System

The limbic system is basically responsible for controlling emotions and learning and memory functions. Located in the structures of the brain, it has the shape of a greyish cortical ring, formed by neurons, and has various structures, each with specific functions. Its structures include the Hypothalamus, which is the size of a pea and represents less than 1% of the brain's size. Its function is to regulate sleep, libido, appetite and body temperature, cooling the blood. The mammillary bodies are related to the hypothalamus and have the function of regulating eating reflexes, swallowing and the desire for certain foods. The Thalamus is found on both sides of the brain and is responsible for man's four senses. It also regulates the sensations of pain, hot and cold, and the pressure of an environment in the ears. The Cingulate Gyrus is another structure adjacent to the Thalamus which, when stimulated by drugs or psychological problems, can cause hallucinations, changes in emotions, as well as controlling smell and vision. The Amygdala is duplicated in the brain, located in the temporal lobe on both sides, and is responsible for feelings of danger, fear and anxiety. And last but not least, the Hippocampus, located in the temporal lobe, is responsible for recent memory, acting when the individual has access to recent memories, causing the metabolism to increase blood flow (https://www.portaleducacao.com.br/medicina/artigos/50288/conhecendo-as-   funcoes-do-sistema-limbico. 2017).

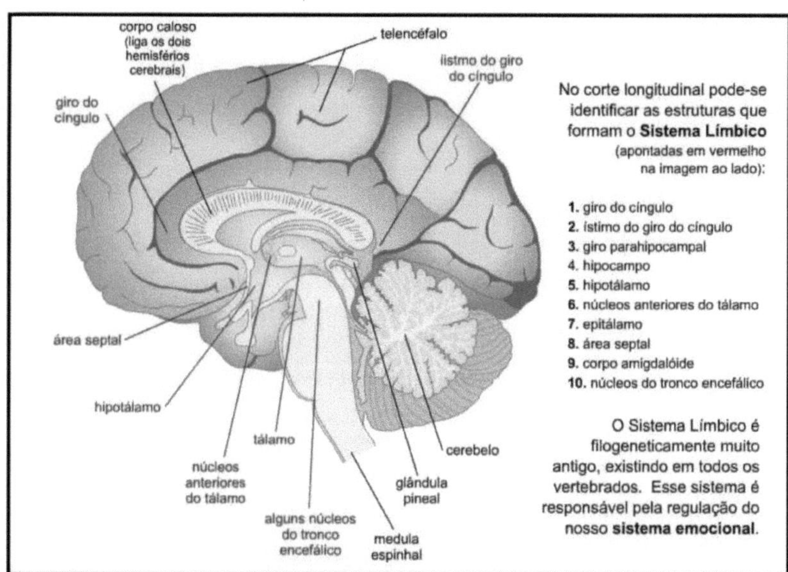

Figure 2: Limbic system. Main structures of the limbic system.

Source : (maissaudeemocional.blogspot.com. 2017).

The limbic system is the emotional centre of the brain. It triggers the feeling of reaction to stimuli, whether real or imagined. Functional magnetic resonance imaging (fMRI) shows which brain areas are activated and proves that psychopaths lack feelings (http://www.reidese.com.br/artigos/032011/03201 1_2.pdf).

Many authors over the years have demonstrated the central role of the limbic system in forming and experiencing emotions, including the mother-child bond, friendships and partner affiliations. Recent studies have gone further to implicate two structures closely related to the limbic system, the insula and the anterior cingulate cortex (ACC), to be central in experiencing and evaluating the emotions of the self and others. These findings are significant because they go beyond the neurobiology of emotions to explain the neurobiology of empathy. The discovery of mirror neuron pathways (activation of areas

The observation of the brain's motor behaviour when performing a task for oneself and when observing it being performed by another) was central in defining theories about neural pathways of empathy. Firstly, this observation was extrapolated to the hypothesis that the mirror-neuron mechanism allows us to identify emotions such as fear, anger and disgust in others, like ourselves, and experience them. Secondly, it was assumed that in callous individuals (with some deformity), these pathways are abnormal compared to the "normal population" (CH ATURA KA et al., 2010).

Neuroradiologist Jorge Moll and neuropsychiatrist Ricardo de Oliveira have developed the "Moral Emotions Battery" (BEM), which uses the technique, still being tested, of fMRI to demonstrate that the brains of psychopaths don't behave in the same way as ordinary people when they are exposed to different images: the psychopath's brain reacts in the same way to both pleasant and perverse scenes. Fear, anger and disgust would be natural reactions to unpleasant scenes. Ordinary people would be agitated by disgusting scenes and serene when they see pleasant ones. However, this doesn't happen with psychopaths, showing that they have reduced brain activity in the areas related to emotions. Rather, there is an increase in activity in the regions responsible for cognition (the ability to reason). It can therefore be concluded that psychopaths are much more rational than emotional, and that their limbic system is hypofunctional.
(http://www.reidese.com.br/artigos/032011/03201 1_2.pdf).

It is believed that reduced limbic and paralimbic circuit activity should affect a person's ability to appreciate the emotions of others (especially fear), engage in appropriate pro-social behaviour (helping, comforting, altruism) and avoid activities that cause discomfort to others. At the same time, the individual may have difficulty processing their own emotions, assessing self-vulnerability and reducing behaviours that put themselves at risk.

A historical case that took place in the middle of the 19th century in Vermont, USA, clearly demonstrated the close association between moral behaviour and brain damage (SILVA. 2008).

Phineas Gage works on a railway. He was well-liked by everyone, a good worker and a great family man. In 1848, an explosion at work caused an iron bar to pierce his brain in the region known as the prefrontal cortex. Amazingly, Gage did not lose consciousness and survived the injury without any apparent sequelae. He could walk normally and his memories were preserved. However, as time went by, Gage became someone else: indifferent in affect, prone to fits of rage and lacking any politeness towards the people around him. Gage was never again the man everyone admired, the "pre-accident" man. Although he never murdered anyone, his life was a pathetic succession of underemployment, fights, drunkenness and petty scams (SILVA. 2008).

The story above played a decisive role in the study of human behaviour, as it was living proof that changes in moral sense can occur when the brain suffers damage to specific areas (in this case, the prefrontal lobe). From this episode onwards, scientists began to research the cerebral roots of amoral behaviour (SILVA. 2008).

CHAPTER 10

Genetic impulsivity

Impulsivity is the tendency to act without foresight. It has been associated with many psychiatric disorders, including addictions, attention deficit hyperactivity disorder (ADHD), bipolar disorder and personality disorders, as well as borderline personality disorder (BPD) and antisocial personality disorder (ASPD). The fourth edition of the *Diagnostic and Statistical Manual* (text revision) also identifies a group of psychiatric illnesses that are collectively defined as impulse control disorders not classified elsewhere. These include intermittent explosive disorder (IED), pyromania, kleptomania, pathological gambling and trichotillomania. Finally, impulsivity is associated with suicidal behaviour, aggression and certain forms of criminality (BEVILACQUA; GOLDMAN. 2013).

Pharmacological studies have implicated various neurotransmitters in impulsivity and several associated genes, altering the function of these neurotransmitters. Dopamine and serotonin-releasing neurons are prominent in brain regions that regulate impulse control. Dysregulated monoamine neurotransmitter activity has been shown to be involved in impulsivity in neuropharmacological, gene knockout and genetic association studies. Some genes alter the function of monoamine neurotransmitters, which have been associated with impulsivity and aggression (BEVILACQUA; GOLDMAN. 2013).

Serotonin is the molecule that has been most consistently associated with impulsivity, namely in what manifests as impulsive aggression and suicide. It has been associated with aggression and impulsivity by neurochemical and neurobehavioural studies in humans and animal models. Tryptophan hydroxylase 2 (***TPH2* gene**) codes for the enzyme that catalyses the rate-limiting step for serotonin biosynthesis in the brain (BEVILACQUA; GOLDMAN. 2013).

The polymorphism of enzyme 2 (TPH2) is the main limiting enzyme in the serotonin synthesis pathway. This polymorphism can be involved in a variety of behavioural phenomena, including mood disorders and suicide. The TPH2 enzyme modifies tryptophan, an essential amino acid found mainly in animal foods, into 5-hydroxytryptamine (serotonin).

As these data are important for the diagnosis and treatment of depression, some authors have already worked on this subject, suggesting that the levels of serotonin and tryptophan in individuals depend, among other things, on their dietary intake, thus creating an analogy between food and depression (BEVILACQUA; GOLDMAN. 2013).

CHAPTER 11

GENES

A gene is the basic physical and functional unit of heredity. Genes, which are made up of DNA, act as instructions for making molecules called proteins. In humans, genes vary in size from a few hundred DNA bases to more than 2 million bases. The Human Genome Project has estimated that human beings have between 20,000 and 25,000 genes. Each person has two copies of each gene, one inherited from the father and one from the mother. Most genes are the same in all people, but a small number of genes (less than 1 per cent of the total) are slightly different between people. Alleles are forms of the same gene with small differences in their sequence of DNA bases. These small differences contribute to each person's unique physical characteristics (GENETICS HOME REFERENCE, 2017).

11.1 HTR2B

Cognitive impulsivity and actual impulsive behaviour show large inter-individual differences. Impulsivity can improve performance in some areas of life, but it can also be a diagnostic feature of neuropsychiatric disorders. Several distinct neural pathways contribute to the complex construction of impulsivity (TIKKANEN et al., 2015).

Bevilacqua *et al.* discovered a termination codon in the mutation of the gene that codes for the serotonin 2B receptor (*HTR2B* Q20 *), located at 2q36.3-q37.1, in a Finnish population. They observed that the stop codon leads to an interruption in the expression of the serotonin 2B receptor (5-HT2B) in lymphoblastoid cells, which implies a 50 per cent decrease in the production of the receptor protein in heterozygous individuals. The true function of the 5-HT2B receptor is poorly understood, especially in humans. However, the researchers found that *HTR2B* Q20 is related to impulsive behaviour and cognitive impulsivity. In addition, they showed that the 5-HT2B receptor is widely expressed in the human brain with the highest densities in the frontal lobe, cerebellum and the occipital lobe, although not all brain areas were examined. This receptor seems to be necessary for the pharmacological anti-depressant action in rats, which indicates a role for the receptors in serotonergic neurotransmissions. It has also been shown in neuronal cells that 3,4-methylenedioxymethamphetamine (MDMA, commonly known as "ecstasy") selectively binds to and activates 5-HT2B receptors, inducing the release of serotonin in the rat raphe nuclei, thus leading to the release of dopamine (TIKKANEN et al., 2015).

This gene codes for one of several different receptors for 5-hydroxytryptamine (serotonin), which belongs to the family of G protein-coupled receptors. Serotonin is a biogenic hormone that functions as a neurotransmitter, a hormone and a mitogen. Serotonin receptors mediate many central and peripheral physiological functions, including the regulation of cardiovascular function and impulsive behaviour. Family-based analyses of a minor allele (glutamine-to-stop substitution, designated Q20) that blocks the expression of this protein, and knockout studies in mice, suggest a role for this gene in impulsivity. However, other factors, such as high testosterone levels, may also be involved in the individual's behaviour. Some alternative splicing transcript variants have been found for this gene (www.ncbi.nlm.nih.gov/gene/3357. 2017).

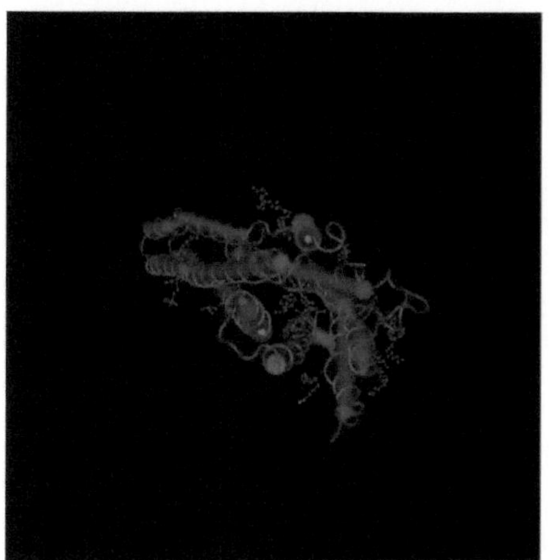

Figura 3. Three-dimensional monomeric structure of 5-HTR2B.

Source: Crystal Structure of the Lsd-bound 5-ht2b Receptor, NCBI. 2017.

The G protein-coupled receptor for 5-hydroxytryptamine (serotonin) also functions as a receptor for various alkaloid derivatives and psychoactive substances. Binding causes a conformational change that triggers signalling via guanine nucleotide-binding proteins (G proteins) and modulates the activity of offspring. Members of the beta-arrestin family inhibit signalling via G proteins and mediate the activation of alternative signalling pathways. Signalling activates a phosphatidylinositol-calcium second messenger system that modulates the activity of the phosphatidylinositol 3-kinase and downstream signalling cascades and promotes the release of Ca (2+) ions from intracellular stores. It plays a role in regulating dopamine and 5-hydroxytryptamine release, 5-hydroxytryptamine uptake and in regulating extracellular dopamine and 5-hydroxytryptamine levels and thus affects neural activity. May play a role in pain perception. Plays a role in regulating behaviour, including impulsive behaviour. It is necessary for the normal proliferation of embryonic cardiac myocytes and normal cardiac development, protects cardiomyocytes against apoptosis, plays a role in the adaptation of pulmonary arteries to chronic hypoxia, plays a role in vasoconstriction. It is essential for normal osteoblastic function and proliferation, and for maintaining normal bone density, and is also necessary for the normal proliferation of interstitial cells of Cajal in the intestine (WEIZMANN INSTITUT OF SCIENCE. 2017).This gene is found in the cytogenetic band of chromosome 2 (q37.1).

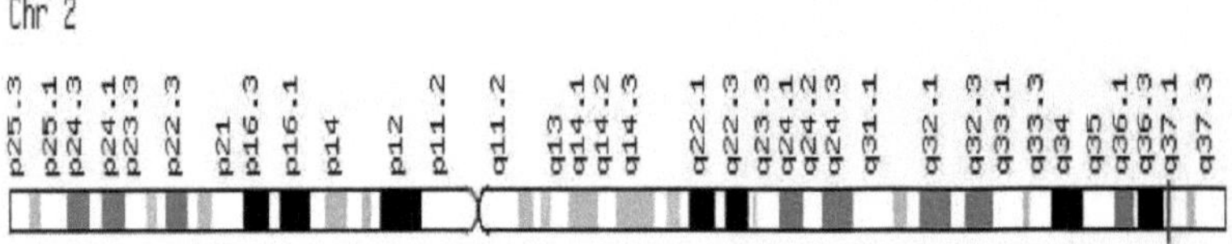

Figure 4 -   Chromosome 2, q37.1. Chromosome 2 is one of the 23 pairs of chromosomes in the human karyotype. It is the second largest human chromosome. It has around 237 million base pairs and represents almost 8% of the total DNA in cells. It contains between 1,300 and 2,000 genes.

Source: Chromosome2human . https://pt.wikipedia.org/wiki/Cromossoma_2_(human). Consulted on 20/08/2017.

112 **MAOA**

Monoamine oxidase A (MAOA) is one of two members of the family of neighbouring genes that code for mitochondrial enzymes that catalyse the oxidative deamination of amines such as dopamine, noradrenaline and serotonin. Mutation of this gene results in Brunner's syndrome. This gene, located on the "X" chromosome, has also been associated with a variety of other psychiatric disorders, including antisocial behaviour (https://www.ncbi.nlm.nih.gov/gene/4128. 2017).

They are found attached to the outer membrane of mitochondria in most types of cells in the body. Two subtypes of MAO have been identified: MAO-A and MAO-B (WEIZMANNINSTITUT OF SCIENCE. 2017).Because MAOA is located on the "X" chromosome (Xpll.4- Xpll.3), all males are homozygous for a single allele, with variants of 2, 3, 3.5, 4 and 5 repeats described. In samples of European descent, the 3- and 4-repeat alleles account for more than 95 per cent of the variations. The 2, 3 and 5 repeat variants are commonly grouped as "low activity" alleles and contrasted with "high activity" alleles of 3.5 or 4 repeats, based on in vitro studies. The functional characterisation of 5 repeats is somewhat controversial, however, despite its rarity it suggests an insignificant effect on the results of the study having differences in a grouping of alleles. In addition, some studies have completely disregarded rare variants, analysing only the alleles of 3 and 4 repeats (BYR; MANUCK. 2014).

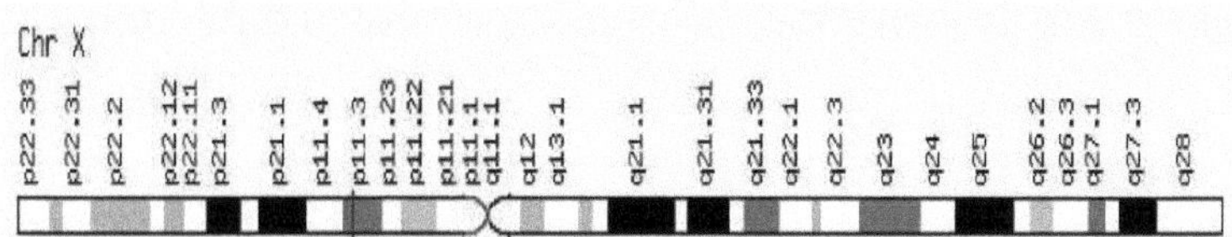

Figure 5 -   **X chromosome.** The X chromosome is one of the chromosomes responsible for determining the sex of human beings. In each set of its 23 pairs of chromosomes, human beings have a pair of chromosomes responsible for sex. Women have two X chromosomes and men have one X chromosome and one Y chromosome.

Source: ChromosomeX   .
https://pt.wikipedia.org/wiki/Cromossoma_X_(human). Consulted on 20/08/2017.

This enzyme is key in the catabolism of monoamines, especially serotonin. Meta-analysis indicates a strong association of the short, less active allele (MAOA-uVNTR-S) with antisocial and criminal behaviour among men exposed to adversity in childhood, which means that large-scale research is needed. Other studies have identified associations between the interaction of the long, high-activity allele (MAOA-uVNTR-L) and childhood adversity and antisocial behaviour among females with aggressive behaviour and violent crime among males. It has been reported that male adolescents carrying MAOA-uVNTR-S showed increased levels of violence when raised in adverse family environments and decreased levels when raised in positive family environments (NILSSON et al., 2015).

Table 1. Genes cards database.

PATHOLOGY	SYMPTOMS
Aggressive behaviour	Aggressive behaviour that can include verbal aggression, physical aggression against objects, physical aggression against people, including aggression against oneself.
Autism	Autism is a neurodevelopmental disorder characterised by impaired social interaction and communication and by restricted and repetitive behaviour. It usually begins in childhood and is characterised by the presence of abnormal development, impaired social development and a restricted repertoire of activities and interests. The manifestations of trastoma vary greatly depending on the level of development and chronological age of the individual.
Behavioural anomaly	Abnormality of mental functioning, including various affective, behavioural, cognitive and perceptual abnormalities.
Cognitive deterioration	Abnormality in the thought process, including the ability to process information.

There is contradictory evidence of epistatic mechanisms being modified by environmental factors. In a study of male offenders, violence was associated with MAOA-uVNTR-S, whereas childhood adversity impacted violence later in life, only when *5-HTTLPR-S* was present (NILSSON et al., 2015).

Exposure to maltreatment in childhood predicted later aggressive and antisocial behaviour among males, as a result of regulatory variation in the gene that codes for monoamine oxidase-A (*MAOA*). The degradative enzyme MAOA deaminates the neurotransmitters serotonin and norepinephrine, and the gene (MAOA) contains a 30 bp repeat sequence (Variable Number of Tandem Repeats) in the 5' flanking region that confers allele-specific variation in the promoter region. In this study, early indicators of maltreatment, such as physical and sexual abuse of boys, maternal rejection or severe physical punishment, later predicted conduct problems, antisocial disposition and violent offending among people carrying the *MAOA* repeat variant (allele) of lower transcription efficiency ("low activity" *MAOA* genotype), than in those of an alternate ("high activity") genotype. This discovery has since been cited more than 2,800 times and prompted similar studies by other researchers. In 2007, Taylor and Kim-Cohen confirmed the interaction of early maltreatment and *MAOA* genotype on antisocial outcomes by meta-analysing the original study and seven attempted replications. All of these studies included male participants recruited from largely normal populations (excluding forensic or predominantly clinical samples) and contained a single or composite index of antisocial behaviour. They also included in the studies a measure of participants' childhood exposure to abuse, neglect or other harm in the family environment, with participants reporting such exposure to be positively associated with study outcomes (BYRD; MANUCK. 2014).

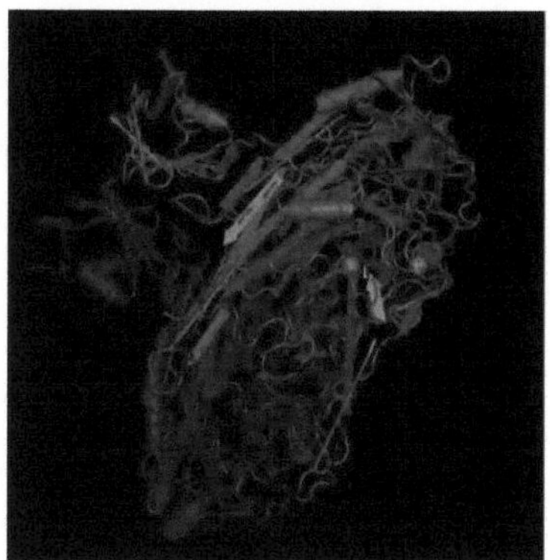

Figura 6. Three-dimensional structure of the MAOA gene. This gene is one of two neighbouring family members that encode mitochondrial enzymes that catalyse the oxidative deamination of amines such as dopamine, norepinephrine and serotonin. Mutation of this gene results in Brunner's syndrome. This gene has also been associated with a variety of other psychiatric disorders, including antisocial behaviour. Alternatively spliced transcription variants encoding multiple isoforms have been observed.

Source: MAOA - Monoamine Oxidase A. https://www.ncbi.nlm.nih.gov/gene/4128. Consulted on 20/08/2017.

CHAPTER 12

Judicial Genetics

In most cases, the only reference to "behavioural genetic information" is the mention of the role of hereditary factors in the aetiology of mental disorders, such as schizophrenia or psychoses, autism spectrum disorders or personality disorders. Experts in most cases (usually a psychiatrist and a psychologist) use information about hereditary factors (and other factors, such as problematic family conditions) only to explain how the disorder and problematic behaviour may have occurred, and the court does not mention this information in its arguments. However, there are also a number of cases in which behavioural genetic information is used to help answer judicial questions (KOGEL; WESTGEEST. 2015).

In one case, the defendant had a hereditary neurological disorder, Huntington's disease, which is known to be caused by a mutated gene on chromosome four. The defendant was accused of setting fire to his girlfriend's house, endangering the lives of several people. Huntington's disease is the cause of neuropsychological problems and dementia that make it difficult for the individual to cope with daily life and problematic situations. In such situations, he reacts with impulsive aggression. The defendant was deemed completely unimputable by the court for his offences. However, the neuropsychological problems and their worsening due to the expected progression of the disease are likely to increase the risk of criminal recidivism, according to the behavioural experts reported to the court. Therefore, the court ordered that the defendant be kept in a psychiatric hospital, a measure that can be continued as long as the person is a risk to others. In this case, the genetic defect and the knowledge about its course and effects on cognitive abilities and behaviour do not lead to any optimism by the experts about the opportunities for reducing the risk of criminal recidivism (KOGEL; WESTGEEST. 2015).In an Italian criminal court case, the influence of the **MAOA gene** on aggressive behaviour played a considerable role. In the first case, a man who had been diagnosed with schizophrenia stabbed another man to death. The defendant had not taken his medication and was psychotic during the offence. He was found guilty by a court of first instance and received a reduced sentence of nine years due to his mental disorder. The appeal court later reduced the sentence to eight years on the basis that the defendant tested positive for genetic variants that made him prone to aggression under stressful circumstances. In another case involving the particular gene (MAOA-gene) mentioned in relation to the treatment of aggressive behaviour, it was the case of a patient who was detained in a so-called "Long Stay Unit" within a forensic psychiatric detention centre, a unit for patients whom the centre considers to have no prospect of successful treatment to reduce their risk of criminal recidivism. The patient attacked three members of staff with a pair of scissors that he had prepared especially for this purpose. The court found him guilty of attempted murder and premeditated serious assault. Behavioural experts, a psychiatrist and a psychologist, reported that the patient has severe antisocial and narcissistic personality disorder and cocaine addiction in remission. They considered the regulation of aggression to be a serious problem in this patient, both in terms of reactive and instrumental aggression. The risk of violent recidivism was seen as strongly elevated. The experts advised placing the patient in a high-security prison unit, since treatment has not led to any improvement. They also advised investigating whether in this scenario problems regulating aggression would be difficult to treat by being influenced pharmacologically. It is in this context that the **MAOA gene** is mentioned. The psychiatrist is quoted as saying: "If it appears that the patient has a reduced expression of the MAOA gene (the so-called 'warrior-gene'), a pharmacological treatment with Depakine could be initiated." The court adopted the expert's opinion and sentenced the patient to nine years in prison. Depakine (valproate) is one of the most prescribed drugs for epilepsy. As valproate has a mood and impulse stabilising effect, it is also prescribed in some cases by psychiatrists to reduce aggressive behaviour (KOGEL; WESTGEEST. 2015).

CHAPTER 13

Treatment

The latest studies into human behaviour reveal that the basic notions of behavioural rectitude and justice depend much less on social learning than psychologists supposed at the beginning of the last century. The latest research into the human brain and comparative analyses of other animal behaviours reveal that the human species acquired the capacity for moral evaluation through natural selection itself. Everything points to the fact that the instructions needed to produce a brain capable of distinguishing right from wrong are already factory-certified, in other words, they are in the DNA of each and every one of us (SILVA. 2008).

Pharmacological therapy is considered ineffective and not recommended in the guidelines for some specific cases. However, it does have a place in the treatment of psychiatric disorders such as depression and anxiety. Given the biological associations of antisocial behaviour (neurotransmitter and hormonal imbalances), the role of pharmacological agents cannot be completely ruled out. One area of interest is the use of selective serotonin reuptake inhibitors (SSRIs). It has been shown that aggression can be linked to dysfunction of the serotonergic nervous system and SSRIs are effective in controlling emotional aggression in personality disorders. However, it has not been shown to be effective in controlling aggression in repeat offenders. Paroxetine (an SSRI) has been shown to improve cooperative behaviour in normal people, but this effect has not been demonstrated in populations with antisocial behaviour. There are case reports of a patient with antisocial behaviour treated with risperidone having his aggression controlled, but it is not mentioned whether he received concurrent psychotherapy. The author attributes risperidone's 5HT (serotonin receptor) antagonism to its therapeutic effect. The observations of this individual case study have not been confirmed by others (CHATURAKA et al, 2010).

The ineffectiveness of pharmacological treatment can be approached using a different hypothesis. In developing an alternative hypothesis on the efficacy (or rather lack thereof) of antidepressants, it is proposed to use a drug-based approach to understanding the effect of antidepressants rather than the traditional disease-based approach. To clarify this, the patient-based approach assumes that the therapeutic efficacy of drugs emanates from their ability to alter the pathology of the disease (for example, SSRIs increase serotonin concentrations that act on synaptic receptors, thus compensating for the "drop" in serotonin that led to depression). The drug-based model proposes that the therapeutic effect of drugs is coincidental and dependent on the social context. Instead of acting on the presumed biochemical model of disease causation, drugs can create a different biochemical environment that can coincidentally alleviate symptoms. He goes further to state that, in such a situation, the effect cannot differ between the placebo and the drug. To support this view, the authors cite the questionable efficacy of antidepressants when prescribed over a longer time scale, the ability of other non-antidepressants to improve scores on depression scales through their sedative/stimulant effects and the conflicting evidence from randomised clinical trials on the efficacy of antidepressants (CHATURAKA et al., 2010).

Discussion

Aggression, lack of emotions and deformities are the combined consequences of genetics, neurotransmitter/hormonal imbalance and environmental factors. Many recent advances have been made in understanding the complex neurocircuitry interrelationships that underpin empathy and emotions. There is intense debate about whether there is a separate diagnosis of psychopathy, but neither antisocial personality disorder as defined in DSM-IV, nor its corresponding diagnosis in ICD-10, dissociative personality disorder, identify psychopathy as a separate diagnosis. A major problem for scholars summarising the evidence on this type of personality disorder is the differences in various diagnostic criteria. The populations diagnosed with these criteria sometimes differ considerably, so a direct comparison of results is difficult. In matters relating to treatment, many psychological and behavioural therapies have shown success rates ranging from 25% to 62% in different situations. Multisystemic therapy and cognitive-behavioural therapy have been shown to be effective in many trials. Given the social and personal costs involved, some authorities, such as the UK's National Health Service (NHS), recommend identifying children at risk and intervening at an early age. This raises several ethical issues that need to be addressed by a wider public discussion.

Genetic pre-disposition or biological vulnerability would be characterised in a child with an emotional deficit. Such a child has a mental system that is deficient in perceiving emotions and feelings, regulating impulsivity and experiencing fear and anxiety. In cases where the parents (family) carry out their educational tasks very competently, these biological characteristics can be compensated for or channelled into socially acceptable activities. However, when the environment is unable to cope with this genetic baggage, whether due to educational failures on the part of the parents, poor socialisation or the fact that this genetic baggage is very marked, the result will be a psychopathic individual (SILVA, A.B.B. 2008).

Conclusion

More in-depth analyses of the interrelationship between neurocircuits, neurotransmitters and hormones in relation to empathy and violence are the consensus among different professional bodies on uniform criteria for the diagnosis of antisocial personality disorder with clarification of the taxonomic existence of "psychopathy", the therapeutic effectiveness of therapeutic communities, family management strategies and contingency management and randomised controlled trials to evaluate the effectiveness of early intervention therapy for "at risk" children, which are identified as areas for future research.When talking about the ills of our world, Albert Einstein once said: "The world is a dangerous place to live, not exactly because of the people who are bad, but because of the people who do nothing about it."

Bibliographic reference

TRISOMIES. **Trisomy XYY.**
http://astrissomias.blogspot.com.br/2011/05/trissomia-xyy.html. Consulted on 20/08/2017.

BARROS, A. J. S. ; TABORDA, J. G. V.; ROSA, G. R. **The Role of Hormones in Psychopathy.** Journal Debates in Psychiatry. Jan/Feb 2015.

BERTOLDI, M. E. et al. **Psychopathy.** Revista da Jornada de Iniciação Científica e de Extensão Universitária do Curso de Direitos das Faculdades Integradas Santa Cruz de Curitiba - ISSN 2357-867X. 2013.

BEVILACQUA, L.; GOLDMAN, D. **Genetics of impulsive behaviour.** Philos Trans R Soc Lond B Biol Sei. 2013 Apr 5; 368(1615): 20120380.

CRIMINOLOGY AND FORENSIC PSYCHOLOGY BLOG. **Serial-killers- parte- viii-explicacoes.html.** https ://psicologia- forense.blogspot. com.br/2014/06/.

BYRD, A. L.; MANUCK, S. B. **MAOA, childhood maltreatment and antisocial behaviour: Meta-analysis of a gene-environment interaction.** Biol Psychiatry. 2014 Jan 1; 75 (1): 10.1016 /j.biopsych.2013.05.004.
CALLEGARO, M. M. **Neurobiology and evolution of psychopathy.** Journal of Psychology. 2010. www.cesusc.edu.br.
CHATURAKA, R.; SENAKA, R.; GAMINI, J. **The 'antisocial' person: an insight in to biology, classification and current evidence on treatment.** Ann Gen Psychiatry . 2010; 9: 31.

COLLECTION OF ARTICLES AND TRUE CRIME AND JUSTICE. **True Crimes XL.** http://truecrimecases.blogspot.com.br/2012/08/bobby-joe- long.html.
CROCKETT, Molly et al. **Dissociable Effects of Serotonin and Dopamine on the Valuation of Harm in Moral Decision Making.** Journal of Current Biology. Curr Biol. 2015 July 20;25(14):1852-1859. Doi: 10.1016/j.cub.2015.05.021.

ESTEVINHO, F. M. ; FORTUNATO, J. M. S. **Dopamine and Receptors.** Revista Portuguesa de Psicossomática, vol. 5, núm. 1, June, 2003, pp. 21-31.

GENETIC LITERACY PROJECT. **Genes-linked-to-violent-crime-but-can- they-explain-criminal-behaviour/.**
https://www.geneticliteracyproject.org/2014/10/29.

KOGEL, C. H. ; WESTGEEST, E. J. M. C. **Neuroscientific and behavioural genetic information in criminal cases in the Netherlands.** J Direito Biosci. 2015 Nov; 2 (3): 580-605.

LEITE, G. **Brief account of the history of criminology.** http://ambito-juridico.com.br/site/index.php7n link=revista artigo leitura&artigo id=6341.

MELDAU, D.C. **Serotonin.** http://www.infoescola.com/neurologia/serotonina.

NATIONAL CENTRE FOR BIOTECHNOLOGY INFORMATION. **HTR2B 5-hydroxytryptamine receptor 2B** *Homo sapiens* **(human),** https ://www.ncbi.nlm.nih.gov/gene/3357.

NATIONAL CENTRE FOR BIOTECHNOLOGY INFORMATION. **MAOA. Human Monoamia Oxidase A,** https://www.ncbi.nlm.nih.gov/gene/4128. Consulted on 20/08/2017.

NILSSON, K. W. et al. **Genotypes Do Not Confer Risk For Delinquency but Rather Alter Susceptibility to Positive and Negative Environmental Factors: Gene-Environment Interactions of BDNF Val66Met, 5-HTTLPR, and MAOA-uVNTR.** Int. J. Neuropsychopharmacol. 2015 Mar; 18 (5): pyul07.
NORDQUIST, N.; ORELAND, L. **Serotonin, genetic variability, behaviour, and psychiatric disorders - a review.** Ups J. Med. Sei .2010 Mar; 115 (1): 2- 10.

PIRES, G. L.; LEITE, M. H. **Common criminals or psychopaths?** http://www.reidese.com.br/artigos/032011/03201 1_2.pdf.

EDUCATION PORTAL. **Getting to know the functions of the Limbic System.** https://www.portaleducacao.com.br/medicina/artigos/50288/conhecendo-as- functions-of-the-limbic-system.

EMOTIONAL HEALTH. **Limbic system.** http://maissaudeemocional.blogspot.com.br/search?q=sistema+l%C3%ADmbic o. Consulted on 19/08/2017.

SECRETARIAT OF EDUCATION OF THE GOVERNMENT OF PARANÁ. **Serotonin.** http://www.quimica.seed.pr.gov.br/modules/galeria/detalhe.php?foto=1942&eve nto=5. Consulted on 19/08/2017.

SEO, D.; PATRICK, J. C.; PATRICK, J. K. **Role of Serotonin and Dopamine System Interactions in the Neurobiology of Impulsive Aggression and its Comorbidity with other Clinicai Disorders.** H.H.S. Public Access. 2008 Oct; 13 (5): 383-385..

SILVA, A. B. B. **Mentes Perigosas: O psicopata mora ao lado.** Editora objetiva Ltda. 2008.

SYMPTOMS AND TIPS. **Genetic Syndromes.** http://www.sintomasedicas.com/sindromes-geneticas/. Consulted on 17/08/2017.

TIKKANEN, R. et al. **Impulsive alcohol-related risk-behaviour and emotional dysregulation among individuals with a serotonin 2B receptor stop codon.**
 Trad Psychiatry. 2015 Nov; 5(1 1):e681. Published online 2015 November 17. doi: 10.1038/tp.2015.170.

 U.S. National Library of Medicine. **Genetics Home Reference.** https://ghr.nlm.nih.gov/primer/basics/gene. 2017. Consulted on 17/08/2017.

WACKER, D. et al. **Crystal Structure of an LSD-Bound Human Serotonin Receptor.** N.C.B.I. *Cell* (2017) **168** p.377-389.

WIKIPEDIA. **Human chromosome 2.**
https://pt.wikipedia.org/wiki/Cromossoma_2_(human). Consulted on 20/08/2017.

WEIZMANNINSTITUT OF SCIENCE. **Gene Card Human Gene Database.** http://www.genecards.org/cgi-bin/carddisp.pl?gene=MAOA&keywords=MAOA.

Buy your books fast and straightforward online - at one of world's fastest growing online book stores! Environmentally sound due to Print-on-Demand technologies.

Buy your books online at
www.morebooks.shop

Kaufen Sie Ihre Bücher schnell und unkompliziert online – auf einer der am schnellsten wachsenden Buchhandelsplattformen weltweit! Dank Print-On-Demand umwelt- und ressourcenschonend produziert.

Bücher schneller online kaufen
www.morebooks.shop